# Quantum Physics

*Beginner's Guide to the Most Amazing Physics Theories*
2nd Edition

**Jared Hendricks**

© **Copyright 2015 by Jared Hendricks- All rights reserved.**

This document is geared towards providing exact and reliable information in regards to the topic and issue covered. The publication is sold with the idea that the publisher is not required to render accounting, officially permitted, or otherwise, qualified services. If advice is necessary, legal or professional, a practiced individual in the profession should be ordered.

- From a Declaration of Principles which was accepted and approved equally by a Committee of the American Bar Association and a Committee of Publishers and Associations.

In no way is it legal to reproduce, duplicate, or transmit any part of this document in either electronic means or in printed format. Recording of this publication is strictly prohibited and any storage of this document is not allowed unless with written permission from the publisher. All rights reserved.

The information provided herein is stated to be truthful and consistent, in that any liability, in terms of inattention or otherwise, by any usage or abuse of any policies, processes, or directions contained within is the solitary and utter responsibility of the recipient reader. Under no circumstances will any legal responsibility or blame be held against the publisher for any reparation, damages, or monetary loss due to the information herein, either directly or indirectly.

Respective authors own all copyrights not held by the publisher.

The information herein is offered for informational purposes solely, and is universal as so. The presentation of the information is without contract or any type of guarantee assurance.

The trademarks that are used are without any consent, and the publication of the trademark is without permission or backing by the trademark owner. All trademarks and brands within this book are for clarifying purposes only and are the owned by the owners themselves, not affiliated with this document.

# Table of Contents

Introduction .................................................................................. 1

Chapter 1 – Quantum Physics: The Beginning ............................... 3

Chapter 2 – Wave Particle Duality .................................................. 6

Chapter 3 – Quantum Tunnelling .................................................. 11

Chapter 4 – Quantum Entanglements, Quantum Optics and Electrodynamics (QED) ................................................................ 14

Chapter 5 – Unified Field Theory .................................................. 19

Chapter 6 – Black Body Radiation ................................................ 22

Chapter 7 – Photoelectric Effect ................................................... 26

Chapter 8 – Young Double Slit Experiment .................................. 30

Chapter 9 – De Broglie's Hypothesis ............................................ 33

Chapter 10 – Compton Effect ....................................................... 35

Chapter 11 – Heisenberg Uncertainty Principle ............................ 38

Chapter 12 – Causality in Quantum Physics ................................ 41

Chapter 13 – Superstrings Theory ................................................ 45

Chapter 14 – Hidden Dimensions ................................................. 50

Chapter 15 – Bohr-Einstein Debates ............................................ 52

Chapter 16 – Physics in the Real World ....................................... 56

Conclusion ................................................................................... 58

# Introduction

Have you ever wondered how scientists produce their explanations about light, energy and matter on molecular level? How can those same scientists measure something they cannot even see? After all, the molecular level is hardly visible to the naked eye. Quantum Physics is the study of the behavior of matter and energy on a molecular level. Think of the smallest particles we know about, such as atoms, protons, neutrons and electrons. These are the building blocks of all living things and are the smallest parts of matter and energy. When studying them, mathematics is the key to really understanding how these small parts of the world work together on a larger scale.

When using these mathematical equations, scientists find the constants within the physical laws on the molecular level and plug these constants into their equations to better understand how these physical laws act on matter and energy. Understanding how matter and energy behave allows for other real life applications to come into play.

In addition, scientists use these mathematical equations to explain what they observe in the world around them and also what they observe through various experiments. As the tools of their trade have become more precise, scientists are able to gather better information to add to their understanding of the molecular world. Today, we benefit from the work of these scientists to better understand our world and the Universe on a molecular level. As we will see, Quantum Physics is mathematics at work explaining the world around us, down to the smallest detail.

Quantum Physics has been defined by its history and the various theories this molecular study has spawn. These theories include wave particle duality and quantum tunnelling. Yet before the scientists could create these theories, there were plenty of experiments which assisted them in formulating these theories.

The experiments included a black body radiation experiment, whose observable results confounded scientists, until one researcher came up with an equation that matched the data they were observing. Other theories, such as the photoelectric effect, was the beginning of a run of experiments and hypothesis that challenged the classic wave theory. Over time, these hypotheses and experiments have built the foundation of data that is the basis for quantum physics or quantum mechanics. The two terms can be used interchangeably and we do so as this book unfolds.

The experiments discussed include the Double Slit Experiment and how it effects the Classic Wave Theory. At the same time, these experiments gave scientists the chance to observe effects that would contribute to the theories that are now part of quantum physics. Other theories highlighted within these pages include the Photoelectric Effect, the Compton Effect and even the uncertainty principle.

Throughout this book, we'll explore some of these experiments and theories, both how they came to be and then how they have grown to become critical parts of what we now know as quantum physics.

# Chapter 1 – Quantum Physics: The Beginning

The Earth and the Universe, in particular matter and energy that are their building blocks, are governed according to the various laws of physics. No matter where we go or what we do, these physical laws are always in force and remain absolute. These physical processes govern how matter and energy can be transformed and its behavior in various situations where they interact with other elements or forces. Yet beyond the physical aspects of the world we can see, there is another microscopic world operating under its own set of laws, also governing the behavior of matter and energy. Scientists describe this set of laws in a group of theories known as Quantum Physics, or the study of how matter and energy behave on the atomic, nuclear and even smaller microscopic levels.

Quantum is a Latin word meaning "how much". In Quantum Physics, the quantum describes the various discrete or distinct units of energy and matter that are predicted by or observed on a microscopic level. This field of study began as scientists gained the technological tools to measure the world even more precisely. The beginning of quantum physics, as a field of study, has been attributed to Max Planck's 1900 paper on blackbody radiation. Development within the field was done by various scientists, including Max Planck, Albert Einstein, Niels Bohr, Werner

Heisenberg, Erwin Schrodinger and others. Let's meet Max Planck and see how his work really opened up Quantum Physics to the scientific community.

*The Father of Quantum Physics*

In 1874, Max Planck, a scientist who had conducted experimental research in the diffusion of hydrogen through heated platinum before turning to theoretical physics, turned his attention to the ultraviolet catastrophe. This problem was based around the Raymond-Jeans formula, which was used to measure thermal radiation, which is the type of electromagnetic radiation emitted by objects because of their temperature. However, the Raymond-Jeans formula was not successful at actually predicting the results of various experiments. By 1900, this formula had created serious problems for classical physics by calling into question the basic concepts of thermodynamics and electromagnetics involved in the equation. Planck reasoned the formula predicted low-wavelength (high-frequency) radiancy too high. Thus, he proposed that if one could limit the high-frequency oscillations in atoms, the corresponding radiancy of high-frequency (again, low-wavelength) waves would also be reduced, which would allow matching experimental results.

Planck suggested that an atom can absorb or reemit energy only in discrete or distinct bundles called quanta. If the energy of these quanta is proportional to the radiation frequency, then at large frequencies the energy would similarly become large. Since no standing wave could have an energy greater than $kT$, this put an effective cap on the high-frequency radiancy, thus solving the ultraviolet catastrophe. While Planck may not have believed quanta was a true physical requirement, but it was a mathematical artifact that helped equations to fit the reality they were measuring.

His work provided a fundamental concept for physics, that energy exists in distinct packets that cannot be broken down any further. For example, Einstein used this concept to explain photoelectric effect in 1905, thus helping to establish the concept of the photon. However, Planck assumed that the Copenhagen interpretation was flawed and eventually, a better theory would replace his concept without the troublesome aspects of quantum theory, such as wave particle duality. Instead, his work and reputation helped to cement the controversial theory of relativity as proposed by Albert Einstein.

*"A new scientific truth does not triumph by convincing its opponents and making them see the light, but rather because its opponents eventually die, and a new generation grows up that is familiar with it." – Max Planck as quoted by philosopher of science Thomas Kuhn in* ***The Structure of Scientific Revolutions***

So what makes Quantum Physics so special within the broader scope of Physics itself? To answer that, it's important to remember that Quantum Physics uses math to explain how energy and matter behave. In other sciences, the observation of an experiment or a phenomenon does not influence the processes taking place. Yet with Quantum Physics, observation does influence the processes, because the equations are developed to explain what was observed. As the next few theories display, it's the scientists' observations that guide the overall development and the adjustments of the mathematical equations that are the brains of Quantum Physics.

# Chapter 2 –
# Wave Particle Duality

Throughout history, science has been fascinated with light and how it behaves. Prisms, among other tools, have been used to observe and measure light. During the 1600s, Christiaan Huygens and Isaac Newton proposed competing theories for light's behavior. Huygens believed that light functioned as a wave, with various lengths. On the other hand, Newton proposed that light didn't behave as a wave, but as a particle. Newton's position in the scientific community of the time helped make his theory dominant, while Huygens dealt with issues of matching observation to his theory.

To understand how these theories differ, one has to understand how waves and particles behave. We'll use light as an example. Across the electromagnetic spectrum, light waves behave in very comparable ways. When a light wave comes across an object, it is either transmitted, reflected, absorbed, refracted, polarized, diffracted, or scattered. What happens to the light wave depends on its wavelength and the structure of the object encountered.

Generally, a wave has to propagate through some type of medium. Huygens defined that medium as luminiferous aether, but today it is known simply as ether. This explanation was accepted in the scientific community, even though there was no concrete proof it existed. During the 1860s, James Clerk Maxwell quantified a set of equations (known as Maxwell's laws or Maxwell's equations) to explain electromagnetic radiation along with visible light as the transmission of waves. He assumed such an ether was the medium of propagation. His predictions with this medium in mind were consistent with his experimental results. However, no such ether was ever located, but instead it remained a mystery.

In 1720, James Bradley completed astronomical observations in stellar aberration. He found that ether, if it existed, would have to be stationary relative the movements of Earth. Throughout the 1800s, many experiments were created to detect the ether or its movements directly, but with no success. The most famous experiment of that era was the Michelson-Morley experiment, an attempt to measure the movement of the Earth through ether. Though often called the Michelson-Morley experiment, it refers to a series of experiments first carried out by Albert Michelson in 1881. Then those experiments were carried out again with superior instruments and equipment at Case Western University in 1887, with assistance from Edward Morley, a chemist.

Light was known to travel through outer space, which scientists believed was a vacuum. One could create a vacuum chamber and shine a light through it. The evidence was clear that light could move through regions without air or other matter. So how could that be? Huygens' ether was the handy substance scientists used to explain how this was possible. The universe, they claimed, was filled with ether, allowing light waves to travel through space and other regions commonly lacking air or any other matter.

Michelson and Morley decided that if the ether did exist, you should be able to measure Earth's orbital rotation through it. Since ether was believed to be unmoving (static except for the vibration) while the Earth was moving quickly, it stood to reason that one could measure ether by its contact with the Earth.

Imagine for a moment holding your hand outside your window, particularly in a car. While it may not be windy, the force of your own motion (courteous of the car) makes it appear windy. Scientists believed ether should have created what would be in effect an ether wind, which would push or hinder the motion of a light wave.

To test this hypothesis, Michelson and Morley designed a scientific device, called the Michelson interferometer, which was meant to split a beam of light, then bounce it off mirrors so that the split beam moved in different directions before finally hitting the same target. The principle at work was based on the idea that if two beams traveled the same distance along different paths through the ether, they should end up moving at different speeds. So when these beams finally hit the target screen, they would be slightly out of phase with each other, creating an observable interference pattern that could be measured. If this experiment had been successful, it would have been the first definitive proof of the existence of this ether.

The results was disappointing, however, because they found absolutely no evidence of the relative motion bias that these two scientists were hoping to observe and measure. No matter which path the split beam of light took, the light always seemed to be moving at precisely the same speed, so there was no interference to measure.

Ether was finally abandoned with the work of Albert Einstein and his theory of wave particle duality. In 1905, Einstein published his paper explaining the photoelectric effect, in which he proposed that light travel in discrete bundles of energy (quantum). The energy contained with a photon was related the frequency of light. As a result, ether was no longer the necessary medium it had once been. But how did this explain the situations when light was observed acting as a wave, and other times when light acted as a particle?

Experiments, such as the quantum variations of the double slit experiment and the Compton Effect, seemed to confirm that light was in fact a particle. But as experiments continued and the evidence mounted, it became clear that light could act as a wave or a particle depending on the parameters of the experiment and when the observations were made.

*Wave Particle Duality in Matter*

The question of whether such duality also showed up in matter was undertaken by the de Broglie hypothesis, which extended Einstein's work to relate the observed wavelength of matter to its momentum. For de Broglie, Einstein's relationship of wavelength to momentum seemed able to determine the wavelength of any matter. His reasoning for choosing momentum over energy is based on the various energy types available to use in the equation, such as total energy, kinetic energy, or total relativistic energy. For photons, it wouldn't matter because all energy is the same in that instance. But matter is different and so momentum was this 1929 Noble Prize winner's choice.

Just like light, it seemed that matter also exhibited both wave and particle properties under precise circumstances. Obviously, massive objects would exhibit very small wavelengths. But for small objects, it is possible to observe the wavelength, as noted in the double slit experiment with electrons.

But what is the significance if light or matter acts as a wave and a particle?

*Significance of Wave Particle Duality*

The major significance of this theory is that all behavior of light and matter can now be explained through the use of a differential equation that represents a wave function, generally in the form of the Schrodinger equation. As a result, describing reality in the form of waves is the heart of quantum mechanics, the mathematical brain of quantum physics.

The most common interpretation of this theory is that the wave function simply represents the probability of locating a given particle at a given point. These probability equations can exhibit other wave-like properties, resulting in a final probabilistic wave function exhibiting these properties also. In other words, the probability of a particle being present in any location is a wave, but the actual physical appearance of that particle isn't a wave at all.

While the mathematics, though complicated, makes accurate predictions, the physical meaning of these equations are much harder to grasp. Explaining what the wave particle duality really means continues to be a key point of debate. So it should come as no surprise that there are multiple interpretations to explain this, but at the same time these interpretations are bound by the same set of wave equations and must explain the same experimental observations. No easy task, as science continues to dig into what this theory means to the real world.

# Chapter 3 –
# Quantum Tunnelling

As a result of the wave-particle duality, it can appear that particles pass through walls. The phenomenon has been well documented and the process is understood within the rules of quantum mechanics.

Quantum tunnelling (or tunneling) is the quantum-mechanical effect of transitioning through a previously-forbidden energy state. Consider rolling a ball up a hill. If the ball is not given enough velocity, then it will not roll over the hill. This makes sense to many of us who have read the tales of Greek mythology, particularly Sisyphus and his endless quest to roll his boulder up the hill.

But in quantum mechanics, objects do not behave like classic objects, but instead exhibit a wave like behavior (as we discussed in Chapter 2). In thinking of a quantum particle, since is it both a wave and a particle, the particle can in theory extend through the hill because of its wave like qualities. Various probability equations can predict the probability of the particle's location and it has the possibility of being detected on the other side of the hill. As a result, it appears to have tunneled through the hill, thus the name quantum tunnelling or tunneling.

In this case, scientists measured electrons escaping from atoms that shouldn't have the necessary energy to do so. In the normal world around us, this would be similar to a child jumping into the air, but somehow clearing a whole house (gravity not withstanding).

Quantum tunneling is possible because of the wave-like nature of matter. As confusing as it may seem, in the world of quantum physics, particles often act likes waves of water rather than billiard balls. To put it simply, an electron doesn't exist in a single place at a single time and with a single energy, but instead exists within a wave of probabilities. As a result, the particle acts more like a wave and appears to flow in a wave like fashion. Probability predicts the various points of a wave or where a particle will be at any given point in time.

"Electrons are described by wave functions that extend smoothly from the inside to the outside of atoms — part of the electron is always outside the atom," explains physicist Manfred Lein of Leibniz Universität Hannover in Germany.

In one recent experiment, researchers used a laser light to subdue the energy barrier that would typically trap an electron inside a helium atom. This laser reduced the overall strength of the barrier so that an electron that wouldn't have the energy required to escape the atom, could instead cheat, so to speak, and tunnel its way through. The researchers found that the electron tunneled through in a very short window of time. They are currently trying to trace the cycle of the electron and determine the exact moment during that cycle that the electron officially left the energy barrier. So how will they measure something so infinitely small?

To measure this, these physicists looked for the photon of light produced when an electron rejoins the atom after making its escape through the tunnel. In some instances, scientists have used a laser to keep the electron away, thus preventing it from recombining with the atom.

While this is the first time scientists have been able to pinpoint when an electron has tunneled through an atom, it won't be the last. Today, technology is providing scientists with ever more accurate tools to help them measure and understand the molecular world. Previously, theoretical calculations could predicted the timing of quantum tunneling, but never before has it been directly measured and with such accuracy.

The findings could help scientists understand other super-fast processes that rely on quantum tunneling, often found within nature. These experiments are just part of a larger attempt to understand how the Earth and the Universe function within the limits of physical laws. But Quantum Physics also has its central principles that help to define the world around us. We will discuss a few in Chapter 4.

# Chapter 4 – Quantum Entanglements, Quantum Optics and Electrodynamics (QED)

Our parents often told us to watch who we associated with, because it would reflect either poorly or positively on us. One of the bedrock principles of Quantum Physics is Quantum entanglement, though it is also highly misunderstood. In short, Quantum entanglement means multiple particles are linked together in a way that means the measurement of one particle's quantum state controls the possible quantum states of the other particles within the linked group. As such, these particles act on one another, much as the friends of our younger selves did. Let's look at this principle a little closer to understand what science has observed.

*The Classic Quantum Entanglement Example*

The classic example of quantum entanglement is called the EPR Paradox. The EPR Paradox (or the *Einstein-Podolsky-Rosen Paradox*) is a thought experiment intended to exhibit the inherent paradox in the early formulations of quantum theory. This thought experiment is among the best-known examples of Quantum entanglement. The paradox involves two particles that are entangled with each other according to quantum mechanics. Under the Copenhagen interpretation of quantum mechanics, each particle is independently in an uncertain state until it is measured, at which point the particle's state becomes certain.

At that exact same moment, the other particle's state also becomes certain. The reason that this is classified as a paradox is based on the fact that it appears the two particles must have communicated at speeds greater than the speed of light, a conflict with Einstein's theory of relativity. This paradox was at the heart of a debate between Albert Einstein and Niels Bohr.

In the more popular Bohm formulation of the EPR Paradox, an unstable spin 0 particle decays into two different particles, Particle A and Particle B, both heading in opposite directions. Because the initial particle had spin 0, the sum of the two new particle spins must equal zero. If Particle A has spin +1/2, then Particle B must have spin -1/2 and vice versa in order to equal zero. According to the Copenhagen interpretation of quantum mechanics, until a measurement is made, neither particle would have a definite state. Both particles are in a superposition of possible states, with an equal probability of having positive or negative spin.

There are two key points within this paradox that make it troubling to scientists.

1. Quantum physics explanations state that until the moment of the measurement, the particles do not have a definite quantum spin, but instead are in a superposition of possible states.
2. Upon measuring the spin of Particle A, we know for sure the value we'll get from measuring the spin of Particle B.

Another words, whatever Particle A's quantum spin is set by a measurement, then Particle B must somehow instantly know what the spin is that it is supposed to take on. As Einstein pointed out, this is a clear violation of his theory of relativity.

Niels Bohr and others defended the standard Copenhagen interpretation of quantum mechanics, as supported by experimental evidence. The explanation is that the wave function which describes the superposition of possible quantum states exists at all points simultaneously. The spin of Particle A and the spin of Particle B are not independent quantities, but are represented by the same term within the equations. The instant the measurement on Particle A is made, the entire wave function collapses into a single state. Therefore, no communication is occurring at the speed of light.

This relationship means that the two particles are entangled. When you measure the spin of Particle A, that measurement has an impact on the possible results you could get when measuring the spin of Particle B. This has been verified by Bell's Theorem.

A fundamental property of quantum theory is that prior to the act of measurement, the particle does not have a definite state, but is in a superposition of all possible states. Imagine for a moment a cat in box with limited oxygen. Because the cat is unobserved, the cat is both dead and alive, since there is no way to definitively say what the cat's state is. Yet, upon opening the box, the cat's state is immediately defined, just as when a particle is measured and its position is clearly defined.

Though this interpretation does mean that the quantum state of every particle in the universe affects the wave function of every other particle, it does so in only mathematically. There is really no sort of experiment which could ever truly discover the effect in one place showing up in another. We have discussed light and waves throughout these chapters, but now it's time to look at the specialized study of light or photons and their interaction with matter.

*Quantum Optics*

Quantum optics is a field of Quantum Physics dealing specifically with the interaction of photons with matter. The theory is that light moves in discrete bundles or photons as represented by Max Planck's ultraviolet catastrophe paper (see Chapter 1). As Quantum Physics developed through the early part of the 20th century by understanding how photons and matter interacted and were interrelated. This was viewed, however, as primarily as a study of matter, not necessarily light.

In 1953, the maser was developed that emitted coherent microwaves. During 1960, the laser made its appearance, known for emitting coherent light. Using these tools, with a focus on light, Quantum Optics was used to describe this specialized field of study.

The findings of Quantum optics support the view of electromagnetic radiation as traveling in both forms, a wave and a particle, as what we have learned as wave particle duality. By using the findings from quantum electrodynamics (QED), it is possible to define quantum optics in the form of the creation and annihilation of photons.

This approach allows the use of certain statistical approaches to analyze the behavior of light, although whether it represents what is physically taking place is a matter of some debate.

Lasers and masers are the most obvious application of quantum optics. Light emitted from these devices is in a coherent state, which means the light resembles a classical sinusoidal wave. In this coherent state, the quantum mechanical wave function and its uncertainty is distributed equally. Laser light is highly ordered, and generally limited to essentially the same energy state, and by default, the same frequency and wave length.

*Quantum Electrodynamics (QED)*

Quantum electrodynamics (QED) is the theory of the interactions of charged particles with an electromagnetic field. These interactions are described mathematically, not just interactions of light with matter but also those of charged particles with one another. QED is a relativistic theory, because Einstein's theory of special relativity is built into each of the equations. Because the behavior of atoms and molecules is principally electromagnetic in nature, all of atomic physics are considered a test laboratory for the theory. Some of the most precise QED tests are experiments dealing with the properties of subatomic particles known as muons. The magnetic moment of this particle type has been shown to agree with the theory to nine significant digits. Agreement of such high accuracy makes QED one of the most successful physical theories so far devised.

The interaction of two charged particles occurs as part of a series of processes building into increasing complexity. In the simplest, only one virtual photon is involved and each process adds virtual photons. The processes correspond to all the possible ways that the particles can interact through the exchange of virtual photons. Each of these can be represented graphically by means of the so-called Feynman diagrams. Besides furnishing an intuitive picture of the process, this type of diagram prescribes precisely how to calculate the variable involved. Each subatomic process becomes computationally more difficult, and there are an infinite number of processes. The QED theory states that the more complex the process, the smaller the probability of its occurrence.

QED is often called a perturbation theory due to the smallness of the fine-structure constant and the resultant decreasing size of the higher-order contributions. This relative simplicity and the success of QED has made it a model within quantum field theories. Additionally, the picture of electromagnetic interactions as the exchange of virtual particles has carried over to the theories of the other fundamental interactions of matter, the strong force, the weak force, and the gravitational force. But with all these theories floating around, how does one fit them together. The Unified Field Theory is an attempt to create a single theoretical framework, as we'll learn about in Chapter 5.

# Chapter 5 – Unified Field Theory

So what is Unified Field Theory? Albert Einstein first coined the term to describe any attempts to unify the fundamental forces of physics, particularly between elementary particles into a single theoretical framework. Einstein himself searched for such a Unified Field Theory, but was not successful. So what brought this about?

In the past, seemingly different interaction fields or forces appeared to have been unified together. For example, James Clerk Maxwell successfully unified electricity and magnetism into electromagnetism in the 1800s. In the 1940s, Quantum electrodynamics translated his electromagnetism into the terms and mathematical equations of Quantum mechanics. During the following decades, physicists successfully unified strong nuclear interaction and weak nuclear interactions, along with Quantum electrodynamics to create the Standard Model of Quantum Physics.

The current problem with a fully unified field theory is in finding a way to incorporate gravity, which is best explained by Einstein's theory of general relativity, with the Standard Model that describes the quantum mechanical nature of other three fundamental interactions. The curvature of space time, fundamental to general relativity, leads to difficulties in the quantum physics representations of the Standard Model.

Some specific theories that attempt to unify quantum physics with general relativity include:

1. Quantum Gravity - Generally is posed that a theoretical entity or a graviton, which is a virtual particle that mediates the gravitational force. This is what distinguishes quantum gravity from certain other unified field theories. In fairness, some theories typically classified as quantum gravity don't necessary require a graviton.

2. String Theory – This uses a model of one-dimensional strings in place of the particles typically used in Quantum Physics. These strings vibrate at specific resonant frequencies. The formulas resulting from string theory predict more than four dimensions, but the dimensions are curled up within the Planck length.

3. Loop Quantum Gravity – This theory seeks to express the modern theory of gravity in a quantized format. The approach involves viewing space time as broken into discrete chunks. It is viewed by many as the well-developed alternative to quantum gravity outside of string theory.

4. Theory of Everything – This theory is a hypothetical single, all-encompassing, coherent theoretical framework of physics explaining and linking together all physical aspects of the universe.

5. Supersymmetry – A theory of particle physics, is a proposed type of space time symmetry relating two basic classes of elementary particles. The first are bosons, which have an integer valued spin. The other is fermions, which have a half-integer spin. A particle from each group associates with each other, creating a superpartner, with a spin differing by a half integer. Perfectly unbroken supersymmetry, in theory, means that each pair of superpartners shares the same mass and internal quantum numbers, in addition to their spin.

As these theories show, the idea of one unifying theory has been difficult to prove and hard to identify. Unified field theory is highly theoretical, and to date there is no absolute evidence that it is possible to unify gravity with all the other forces. Historically, other forces have been combined, and many physicists are willing to devote their lives, careers, and reputations attempting to show that gravity can also be expressed quantum mechanically. The magnitudes of such a discovery, of course, cannot be fully identified until a viable theory is proven by experimental evidence.

# Chapter 6 –
# Black Body Radiation

The wave theory of light was the dominant light theory in the 1800s. This theory was captured by Maxwell's equations and surpassed Newton's corpuscular theory. However the theory was challenged by how it explained thermal radiation, which is an electromagnetic radiation given off by objects based on their temperature. So how could someone test or detect thermal radiation?

Scientists can test for thermal radiation by setting up an apparatus to detect radiation from an object based on specific temperature, represented by $T_1$. Warm bodies give off radiation in all directions, so in order to be able to measure it effectively, shielding must be used so the radiation is examined in the form of a narrow beam.

In order to create this narrow beam, a scientist use a dispersive medium, such as a prism, placed between the body or object emitting the radiation and the radiation detector. This allows the radiation wavelengths to be dispersed at an angle. Then the detector measures a specific range or angle, essentially the narrow beam. This beam is considered a representation of the total intensity of the electromagnetic radiation across all the wavelengths.

So let's define a few key points. One thing to note is that the intensity per unit of a wavelength interval is referred to as radiancy. Calculus notation helps us to reduce various values to zero and create the following equation: $dI = R(\lambda)\, d\lambda$. Using the prism, a scientist can detect $dI$, or the total intensity over all wavelengths, so one can define radiancy for any wavelength by working backwards through the equation. Now let's look at how we can build a database of sorts for wavelength versus radiancy curves.

Scientists typically perform an experiment over and over again, building up a store of data that creates various ranges. When working with these ranges, one can begin to build a better understanding of how much radiation will occur from a specific object, but also how intense it will be at any given temperature.

For instances, one can glean that as the total intensity radiated increases as we increase or decrease the temperature. But when we look at the wavelength with the maximum radiancy, we find that the inverse occurs, that is with that specific wavelength, the intensity will go down as the temperature increases. Thus, as the temperature go up, wavelengths can change their individual radiation intensity, but the overall radiation intensity will continue to increase with the temperature.

So if the temperature is going to down, then the maximum intensity of an individual wavelength will go up, but the overall or total intensity of the object will go down, corresponding with the temperature.

Again, we go back to how to measure something when light reflects off so many things. How do you create the angle, making sure that you are accurately measuring your narrow beam?

A simple way to do this is to stop looking at the light and look at the object that doesn't reflect it. Light does reflect off of objects, but scientists will perform this experiment observing a blackbody, or an object that doesn't reflect any light at all. Otherwise, the experiment runs into a difficulty defining what is being tested.

Performing this experiment requires a box, preferably metal, with a tiny hole. If or when light hits the hole, it enters the box but it won't bounce back out. As a result, the hole, not the box, is the blackbody of the experiment. Any radiation detected outside of the hole is a radiation sample of the amount of radiation in the box. Scientists analyze this information to understand what's happening within the box.

The first thing to be noted is that the metal box is being used to stop the electric field at each wall of the box, creating a node of electromagnetic energy at each of the walls. Thus standing electromagnetic waves are contained within the box.

Second, the number of standing waves with their various wavelengths within a defined range including an equation that takes into account the volume of the box. By analyzing the standing waves and then following this equation, it can be expanding into three dimensions.

Third, classical thermodynamics contributes a basic truth: the radiation in the box is in a thermal equilibrium with the walls of the box at a certain temperature. The radiation within the box is absorbed and reemitted from the walls constantly, creating oscillating within the radiation's frequency. The thermal kinetic energy of these oscillating atoms are simple harmonic oscillators, so the mean kinetic energy equals the mean potential energy. As a result, each wave contributes to the total energy of the radiation within the box.

Fourth, energy density is related to the radiance. Energy density is defined as the energy per unit volume within the relationship. The measurement of this is determined by the amount of radiation passing through a component of surface area with a cavity.

Classic physics as represented by the Rayleigh-Jeans formula failed to predict the actual results of these experiments, primarily due to the fact that classic physics failed to account for shorter wave lengths. At longer wavelengths, the Rayleigh-Jeans formula more closely matched the observed data. This failure was referred to as the ultraviolet catastrophe. In early 1900, this was a big issue, because it called into question such basic concepts as thermodynamics and electromagnetics as part of that equation.

Historically, this is where quantum physics came into play. Simply, Max Planck used quanta to create what would be defined as discrete bundles of energy. Thus the quanta would be proportional to the radiation frequency. With this theory, no standing wave could have more energy than kT, then high radiation frequency would be capped, solving the ultraviolet catastrophe. In the end, frequency describes the energy of each quanta, where a proportional constant.

While this resulted in an equation that fit the data of the experiments perfectly, but it wasn't as attractive as the Rayleigh-Jeans formula. This formula became the starting point of quantum physics as we know it today. Einstein even demarcated it as a central principal of the electromagnetic field, while Planck had originally used it just to solve the issue of one experiment. While it took scientists a while to warm up to what is now known as Planck's Constant, it is now considered a critical part of the quantum physics or quantum mechanics.

This was just one part of the large array of experiments that define quantum physics. Another early experiment in concert with wave particle duality, a challenge that was known as the photoelectric effect.

# Chapter 7 – Photoelectric Effect

This significant challenge appeared in the study of optics during the 1800s. The photoelectric effect tested the classical wave theory of light, which was predominant at that point in time. In coming up with the solution to this dilemma, Einstein would gain a reputation in the physics community, eventually earning the Noble Prize in 1921.

The Photoelectric Effect was first observed in 1839, it wasn't documented until 1887 in a paper by Heinrich Hertz. Basically a light source is incident upon a metallic surface, the surface emits electrons called photoelectrons.

In order to observe this effect, scientists would create a vacuum chamber with photoconductive metal at one end, plus a collector at the other end. By shining a light on the metal, electrons are released that move through the vacuum to the collector. As a result, a current is created in the wires connecting the two ends. This current is measured with an ammeter. When administering a negative voltage potential to the collector, more energy is expended for the electrons to complete their journey, thus initiating a current. When no electrons find their way to the collector, this is called the *stopping potential* $V_s$, which can be used when defining the maximum kinetic energy of the electrons themselves.

Note that not all electrons have this energy, but will have a range of energies based upon the experimental metal's properties. The equation created to calculate the maximum kinetic energy of the particles bumped free of the metal surface at a maximum speed.

Classic Wave Theory explained as energy of electromagnetic radiation being carried within the wave itself. As the wave collided with the metallic surface, the electrons absorb the wave's energy until it exceeds the binding energy, thus releasing that electron from its metal surface.

This classic explanation includes three fundamental predications:

1. The resulting maximum kinetic energy should have a proportional connection with the strength of the radiation involved.
2. This effect will occur with any light, regardless of wavelength or frequency.
3. A delay will be witnessed of seconds between the radiation's contact with metal and its first release of photoelectrons.

Yet the experimental results directly contrasted with these three predications.

1. Maximum kinetic energy of the photoelectrons wasn't effected by the intensity of the light source.
2. The photoelectron effect was not observed below a certain frequency.
3. The delay was not observed between the radiation's metallic contact and the emission of the first photoelectrons.

Since these are the exact opposite of what was predicted by the wave theory and completely counter-intuitive to what the scientists believed would occur. Einstein would publish a paper in 1905 that built on Max Planck's black body radiation theory to explain the photoelectric effect and the contradictions scientists were observing. What he proposed was that radiation energy is not distributed equally over the wave front, but is localized into smaller bundles called photons.

The photon's energy was associated with its frequency, along with its proportionality constant and the speed of light. The proportional constant could also be defined using the wave's length.

According to Einstein, a photoelectron was released from an interaction with a single photon, rather than interacting with the whole wave. The energy transfers instantaneously to a single electron, knocking that electron free if the energy is high enough to break away from the metal's work function. If the energy or frequency is too low, there won't be any electrons released. With excess energy, beyond what is available in the photon, this excess energy will be converted into the photon's kinetic energy.

So what Einstein put forward is that maximum kinetic energy is completely independent of light intensity. He was so confident that he didn't even add the intensity of light into his equation.

When researchers shine twice as much light, they get twice as many photons, but the maximum kinetic energy itself doesn't change unless the energy of the light, instead of its intensity changes. So when does the maximum kinetic energy occur? It results when the least-tightly bound electrons break free. As for the more tightly bound photons, when they are knocked free there is a result of kinetic energy equal to zero. The result was equations that indicate why the low-frequency light couldn't free any electrons, thus producing no photoelectrons.

Since Einstein, other experimentation has been carried out by Robert Millikan, not only confirmed Einstein's theory, but also won Millikan a Nobel Prize in 1923. This experiment and the resulting data helped to crush the classic wave theory, as it was shown that light behaved as both a wave and a particle, commonly known now as the wave particle duality.

But other scientists were studying light and proving the wave theory with their experiments. One such experiment was the Young Double Slit Experiment.

# Chapter 8 –
# Young Double Slit Experiment

Thomas Young appeared to prove that light was a wave with his double slit experiment. The results of his experiment had a profound effect on physics at the time, including the continued search for the allusive ether, otherwise known as the medium of light propagation. Yet, over time it was found that this experiment could be done with any wave medium, including water. For the purposes of this discussion, we are primarily focused on light. So how was this experiment conducted?

During the early 1800s, Thomas Young allowed light to pass through a slit into a barrier, so the light expanded into wave fronts using that slit as the light source. That light was then funneled through another pair of slits in another barrier. Each of these new slits then diffracted light as if they were individual sources of light. The light's impact on an observation screen was then observed and noted. When a researcher only opened one slit, the impact on the observation screen was a greater intensity near the center, which faded as you moved outward.

This could be explained two different ways. The first is a particle interpretation. If light does exist as particles, the intensity of both slits will equal the sum of the two individual slits.

The second is a wave interpretation. Basically, if light exists as waves, light waves will create bands of light and dark, based on the interference under the principle of superposition. The results is the light bands are constructive interference, while the dark bands are destructive interference.

The interference patterns did show up within this experiment, seeming to support light traveling as a wave. This breathed new life into Huygens's wave theory of light. As a result, the search was on for ether, an invisible medium. Ultimately, these experiments to detect either or its effects failed. With Einstein's work in photoelectric effect and relativity, ether was no longer part of the explanation of light's behavior. So the particle theory again took a dominate role in the scientific community. But once it was determined that light moved in discrete quanta, scientists wanted to know how that was possible. So the Double Slit Experiment was expanded.

At the beginning of the 1900s, scientists wanted to know how light was exhibiting wave characteristics, although they traveled in particle bundles of energy. Water acted as both particles and waves, so the feeling was that maybe light was acting in a similar way. The question was how do you test for that without some technologically advanced equipment?

Once it became possible to have a light source emit one photon at a time, scientists could move forward with this experiment to solve the question. In essence, scientists were sending microscopic balls through the slits. The observance screen was then set up that was delicate enough to perceive a single photon. All this set up allowed them to determine whether the interference patterns were still there.

So how would this play out? With a sensitive film set up as the observance screen, the experiment could be run over a defined period of time. The film could then be analyzed for the pattern of light on the screen. When this experiment was performed, the alternating light and dark bands appeared, which seems to be a result of wave interference.

But how could this be? Photons were being emitted individually and could only go through the single slit one at a time. Attempts to answer this question has resulted in many interpretations, from the many worlds interpretation (we'll discuss that later) to the Copenhagen interpretation.

Now perform the experiment again, but make one change. Place a detector to indicate whether a photon clears through a given silt. If the photon passes through one slit, then it cannot clear the second slit at the same time and cause interference to itself. By adding this point of measurement early in the experiment, the wave bands disappear. The uncertainty of position is related in some way to whether or not wave effects manifest themselves on observations sheets. Would conducting the experiment slightly differently produce different results?

Over the years, scientists attempted this experiment with variations in an attempt to explain the disappearance of the wave interference. In 1961, Claus Jönsson did the experiment and had Young's result of interference patterns on the observation screen. In 1974, the single electron could be released, thanks to advances in technology. But when the detector was placed at the slit, again the interference disappeared. Since that time, the experiment has been done with photons, electrons, as well as atoms, but the results quickly became obvious. When you measured the position of the particle at the slit, the wave behavior disappears. While many theories are in existence strictly to explain that, much of it is still conjuncture.

So while this experiment demonstrates light behaving as a wave, it still doesn't help clear up the reason why light is behaving as both a particle and a wave. Hence, wave particle duality continues to exist as part of quantum physics or quantum mechanics. Still other hypothesis regarding the wave behavior of objects were being performed and becoming part of the lexicon of quantum mechanics.

# Chapter 9 –
# De Broglie's Hypothesis

Einstein's photon theory gained acceptance and now scientists wanted to know if this theory would hold true for other material objects could exhibit the behavior similar to a wave. In 1923, Louis de Broglie, a French physicist, made a bold assertion with his doctoral dissertation. Essentially, de Broglie asserted that the relationship of wavelength to momentum could be used to determine the wavelength of any matter, in a relationship with Planck's constant. This was called the de Broglie wavelength. The reason de Broglie chose the momentum equation versus the energy equation is because it would be difficult with matter to define whether the E in the equation should be total energy, total relativistic energy or kinetic energy. While this wouldn't play a part with photons, matter is different.

But by using the momentum equation, scientists could allow a derivation of the de Broglie relationship for frequency but using kinetic energy instead. These relationships use alternatives as well. One of the alternative formulations is expressed in the terms of Dirac's constant along with the angular frequency and wave number. But a hypothesis is just that until experimental data is accumulated to confirm it.

During the late 1920s, Bell Labs physicists Clinton Davisson and Lester Germer worked on an experiment where they shot electrons at a crystalline nickel target. The diffraction pattern that resulted matched the predictions of the de Broglie hypothesis of wavelength.

The Nobel Prize of 1929 went to de Broglie for his hypothesis and then in 1937, Davisson and Germer won for experimental discovery of electron diffraction and by default, the proving of de Broglie's hypothesis.

Quantum variants of the double slit experiment have held true to de Broglie's hypothesis. Even diffraction experiments done in 1999 confirmed the hypothesis. In these experiments, molecules the size of buckeyballs (groups of carbon atoms that make up complex molecules) supported the de Broglie wavelength.

What makes this particular hypothesis and the data which backed it up so important? This hypothesis showed that wave particle duality is not an aberrant behavior that can be attributed only to light, but was a fundamental principle exhibited by both radiation as well as matter. These wave equations could now be used to describe various material behavior, as long as the de Broglie wavelength equation is properly applied. Quantum physics or quantum mechanics took a step forward and this became a building block in this area of study.

Still, this equation, like any other in quantum physics, has its limitations. Although it works for wavelengths on matter of any size, the wave aspects of a macroscopic object are so small that they become unobservable in any way that could be considered useful.

The wave particle duality was now being taken out of the realm of research and being taken out to the nature world. While it didn't explain how a molecule can act as both a particle and a wave, science has become to except this inherent contradiction and move forward.

# Chapter 10 – Compton Effect

As with everything in quantum mechanics, equations are the math created to explain a variety of events on the molecular level. Scientists are always looking for a better way to explain how electrons as expressed by light or other matter are moving and the energy released as well as gained through that movement. One such equation was created by the Compton Effect, otherwise known as the Compton scattering. The Compton Effect was found to be the result of a high energy photon colliding with a target, releasing loosely bound electrons that were found in the outer shell of the molecule or of an atom.

As a result of the collision, scattered radiation experiences a shift in the wavelength that didn't fit into terms of the classical wave theory. Remember the classic wave theory has been taking a beating, so to speak, from these experiments and hypotheses that are focused on how electrons and matter can be moving in terms of particles and waves. This is yet another blow to the classic wave theory. As we have seen with all these experiments, most of them start with the premise of Einstein's photon theory and appear to show support for that theory.

Arthur Holly Compton received a Nobel Prize in 1927, but the effect named after him was originally demonstrated in 1923. So how does this process known as the Compton Effect really work? Simply put, the high energy photon, produced from a gamma ray or an X-ray collides with a defined target that has loosely bound electrons on the outer shell. This photon is known as the incident photon is defined with the following energy $E$ and linear momentum $p$. Within the Compton Effect, the photon gives a portion of its energy away to another almost free electron in the form a kinetic energy, which is to be expected when you have a particle collision.

Scientists have come to understand that total energy and linear momentum must be conserved. When analyzing these energy and momentum relationships in regards to the photon and electron, three equations are the result. These equations include energy, an x-component momentum and a y-component momentum. There are also four variable involved as listed below:

- *Phi* – which is the electron's scattering angle
- *Theta* – which is the photon's scattering angle
- $E_e$ – which is the electron's final energy
- $E'$ – which is the photon's final energy

Suppose we only focus on the energy and the direction of the photon, then we can treat the electrons as a constant, so we can potentially solve the system of equations for the effect. Compton combined several equations and using a few tricks he picked up from algebra to eliminate some variables, he was able to create the two equations that are related because the energy and wavelength are both related in photons. The results were the following equations:

$$1/E' - 1/E = 1/(m_e c^2) * (1 - \cos theta)$$

$$lambda' - lambda = h/(m_e c) * (1 - \cos theta)$$

The Compton Wavelength of the Electron has a value of $2.426 \times 10^{-12}$ m. While this isn't an actual wavelength, it can be used as a proportionality constant to represent the wavelength shift. So why does this particular effect support protons?

In part, this analysis and derivation is based upon a particle perspective. The results have been easy to test for over time. When observing the equation, clearly the entire shift can be measured or quantified in terms of the angle from which the photon is scattered.

Simply put, everything on the right side of the equation is used as a constant. Since experiments have consistently shown this to be the case, thus supporting the photon interpretation of light.

Understanding some of these theories and the experiments behind them are important to having a greater understanding of quantum physics as a whole. However, nature always throws curve balls and so there are effects that cannot be explained through these theories. So how do scientists define the uncertainty inherent in this study of the smallest things on earth known as quantum mechanics? One such way is by the cornerstone of quantum physics, otherwise known as the Heisenberg Uncertainty Principle.

# Chapter 11 – Heisenberg Uncertainty Principle

So what is involved in the Heisenberg Uncertainty Principle? This cornerstone is not deeply understood even though it is such an important part of quantum mechanics. This is a simple overview and not an in depth study of this principle. Yet an overview can help you to understand how it explains the uncertainty at nature's essential levels. With carefully constructed experiments, one can see how this principle helps to manifest this uncertainty in a constrained way, thus limited the effects on our daily lives.

German physicist Werner Heisenberg worked out what became known as the *Heisenberg uncertainty principle*, otherwise known as the *uncertainty principle* or the *Heisenberg principle* in 1927. Heisenberg was endeavoring to build an instinctual model for quantum physics. In the process, he uncovered certain fundamental relationships that limited how well a scientist could know the quantities they were working with. Simply put, the more precisely a particle's position is measured, the less precisely the particle's momentum can be known.

A variety of relationships came out of Heisenberg's work. His uncertainty principle is a precise mathematical equation or statement about the overall nature of the quantum system itself. As a result, it constrains the degree of precision that can be discussed within a system based on the physical and mathematical terms.

The most common equations related to the uncertainty principle are referred to as the Heisenberg uncertainty relationships, as represented by the following equations:

a) delta-$x$ * delta-$p$ is relative to $h$-bar

b) delta-$E$ * delta-$t$ is relative to $h$-bar

The key to the symbols in the equations are defined in the following way:

- $h$-bar – This is the reduced Planck constant, basically the Planck constant value divided by 2 times pi..
- delta-$x$: This is representative of the uncertainty of an object's position, for example the position of a particle
- delta-$p$: This is representative of the uncertainty in an object's momentum
- delta-$E$: This is representative of the uncertainty in an object's energy
- delta-$t$: This is representative of the uncertainty in the object's time measurement

Based on these equations, a scientist can discern some physical properties of a system's uncertainty measurement based upon its corresponding level of precision with their measurement. It's a long explanation, but with uncertainty proportion is truly the key to understanding it. For example, if uncertainty in any of the measurements gets extremely tiny, this corresponds to an extremely precise measurement and so these relationship equations explain that the matching uncertainty will have to increase in order to maintain proportionality.

In other words, a scientist can't measure both properties within the equation at exactly the same time with a limitless degree of accuracy. So the more precisely we measure one position, the less precisely we can measure the momentum and same is true if we are more precisely measuring the momentum. Additionally, the more precisely a scientist measures times, the less precisely they are able to exactly measure energy. The same would be true in the reverse as well. It is simply impossible to precisely measure two of these items at exactly the same time or in a simultaneous fashion.

The uncertainty relationships essentially are created out of objects' wave-like behavior at the quantum scale and the fundamental difficulty of capturing a precise measurement of the wave's physical position, even in what is known as a classical case.

What causes confusion with the uncertainty principle is the observer effect that is found in quantum mechanics. However, it has to be noted that these are two entirely different issues within the realm of quantum physics. By using the uncertainty principle is a fundamental constraint on how precise our statements can be about the behavior within a quantum system, regardless of whether we observed something or not. With the observation effect, it is implied that if we make an observation, the system itself behaves in a different fashion than if we hadn't made the observation at all.

That really leads into the cause and effect aspects of quantum mechanics and some of the thought experiments related to this field of research.

# Chapter 12 –
# Causality in Quantum Physics

**Part A: Schrödinger's Cat**

The first thought experiment up for discussion is the one called Schrödinger's Cat. Erwin Schrodinger, a scientist who create the defining equation to explain motion in the universe, but that motion was expressed in probabilities. Since most scientists prefer hard facts versus probabilities, including Schrodinger himself, he decided to come up with an illustration to help others understand the issues inherent in quantum physics. Using analogies, Schrodinger came up with the Schrodinger Cat thought experiment. Let's look at a few of the issues Schrodinger attempted to explain with his thought experiment.

One such issue was quantum indeterminacy, where a particle or an atom can be in two different states, at least until it is measured. At that point, its physical reality is determined by the act of measurement. Therefore, any particle or atom remains in a superposition within two quantum states until the time of measurement, when they collapse into one state.

As a result, the observation is what solidifies the physical state one way or another, but without that observation the physical world merely exists in a realm of possibilities. Schrodinger explained this best with his cat in the box. Now within this box is a cat and a vial of poison gas that could kill the cat. Attaching the vial to a Geiger counter and the counter to a radioactive atom completes the players in this analogy. Now the atom itself will decay, registering radiation and breaking the vial, thus killing the cat, or the atom won't decay and the vial will remain unbroken. Because we can't observe the cat and the poison within the box, Schrodinger said this illustrated a particle in two states, because the cat was both dead and alive. Until the box was opened and then the physical state would be defined by the observation.

While many scientists have different interpretations of the thought experiment, the biggest issue appears to be a matter of scale. Simply put, quantum mechanics deals with microscopic particles, not the macroscopic scale of animals and vials. Another objection is that the act of measurement has been done many times before the cat even entered the box, making it nearly impossible to isolate the cat or any of the other parts of the experiment. As a result, they believe that opening the box is irrelevant because the cat is already either dead or alive, but not both.

**Part B: EPR Paradox**

The EPR Paradox, otherwise known as the *Einstein-Podolsky-Rosen Paradox*, is intended to demonstrate some of the paradoxes inherent in early formulas of quantum mechanics. An example is quantum entanglement where two particles are tangled with each other. Each individual particle is not defined until it is measured. Then its state is defined and by default, so is the other particle it is tangled with. The reason a paradox appears is because it seems that the particles must have communicated at a speed greater than the speed of light, and that is a direct conflict with Einstein's own theory of relativity.

This paradox originated as a focus within an intense debate between Einstein and Niels Bohr. Einstein, who didn't agree with many points made by Bohr and his colleagues, created the EPR Paradox with his colleagues Boris Podolsky and Nathan Rosen. This was a way to show that quantum theory was inconsistent with the other physics laws as they were known at that time.

The paradox is based on a particle that is unstable with a quantum spin of zero and eventually decays into two particles. Each of the new particles' spins must equal zero. So one particle that is measured with a spin -1/2, then the other must be a +1/2 in order to equal zero. But until one is measured they both lack a definitive state but have an equal probability of being the negative or the positive.

Here's what made this troubling to the scientists who then pointed it out as a paradox. One, quantum mechanics says until a moment of measurement, a particle does not have definite quantum spin until it is measured. Two, that as soon as one particle's spin is measured, the value is set before we measure the spin of the other particle.

Einstein saw this as a clear violation of the theory of relativity. Instead, he and David Bohm supported an alternative approach, otherwise known as the hidden variables theory. It suggested that quantum mechanics in its current form was incomplete. The missing, but not immediately obvious, needed to be added to explain the non-local effect, as demonstrated by the two particles.

The uncertainty in quantum mechanics isn't just based on a lack of understanding and knowledge, but a lack of a definite reality. The problem is that the hidden variables are hard to find and scientists struggled see how they could be incorporated in a meaningful way.

While Bohr defended quantum theory with the Copenhagen interpretation, which says that the superposition exists simultaneously at all states, therefore explaining the apparent communication between particles because they are represented by the same term with the equations.

The Bell's Theorem was a defining moment against the idea of hidden variables. Again and again, these inequalities were violated and thus quantum entanglement was shown to take place. Today, most professional scientists do not support the idea of hidden variables as put forward through the though experiment of EPR paradox. Our final though experiment is a unique interpretation.

## Part C: The Many Worlds Interpretation

The many worlds interpretation (MWI) used to explain how the universe contains some non-deterministic events while the theory itself is meant to be completely deterministic.

According to this interpretation, each random event splits the universe into the various choices available. Each version contains a different outcome than the others. Imagine a tree with branches splitting off of it. It simply doesn't tell us precisely when a given event will occur.

According to customary quantum theory, until the measurement is made there is no way to know if it has decayed or not. You would have to treat that atom as if it is in a state of superposition, another words, both decayed and not decayed. As we have seen in the Schrödinger's Cat experiment, these contradictions are inherent when applying quantum theory literally.

If quantum theory says an atom is in both states, then MWI concludes two universes must exist, one where the atom is decayed and another where the atom is not. This continues indefinitely, creating an unlimited number of quantum universes. So the Everett Postulate, as part of the MWI, put it forward that the entire universe continuously exists in the multiple states of superposition. Thus there really is no point where the wave function collapses because that would mean that the principles of quantum mechanics weren't being followed by the universe they were created to describe.

Non-physicists use terms such as multiverse, megaverse, or parallel universes to describe the many worlds interpretation. In science fiction, the parallel universes have been used to create some incredible stories, but they aren't based in fact. The MWI doesn't allow for any communication between these parallel universes, which would make most science fiction stories implausible at best.

Now let's leave the debates of parallel universes and head back to electromagnetic fields, in the form of the Superstrings Theory.

# Chapter 13 – Superstrings Theory

Closing the gap between several quantum physics theories is the superstrings theory. But how did it come about? First, let's explore some of the theories and how their gaps allowed for the superstrings theory to be born.

It all starts with light. Electromagnetic fields are defined with mathematics. Then with the discovery of electrons, particle physics was born. As a result of quantum mechanics, both the equations and the observations, particles were divided into two classes, bosons and fermions. Only one fermion can occupy a certain state at a certain time, thus making them the particles of matter. So that's why solids aren't able to pass through each other. Bosons, on the other hand, can occupy the same state at exactly the same time. The Pauli Repulsion explains this inability of matter to share the same space as forces can.

Throughout the development of quantum mechanics, evidence grew up that indicated light always traveled at one fixed speed, no matter the direction. Einstein developed a Special Theory of Relativity to describe this discovery. This theory, along with other developments of quantum physics, resulted in the rich subject known as relativistic quantum field theory. This is the foundation of what physicists use to define the actions of subatomic particles.

Einstein was a busy scientist though. He extended his special theory to encompass Newton's theory of gravitation. When he did that, he defined the General Theory of Relativity and the mathematics of differential geometry into the world of physics.

Relativistic quantum field theory works when we neglect gravity, because it is so weak. Particle theory also seems to work best when scientists pretend that gravity doesn't exist. General relativity has given scientists greater insight into orbits of planets, creation and lives of stars, even black holes. But this theory works only when scientists essentially pretend the Universe is classical and describing nature doesn't require quantum physics. String theory attempted to bridge the gap.

String theory or superstrings theory started out as an explanation for the relationship between spin and mass for hadron particles, which include a proton and a neutron. While another better explanation was found, string theory found a home in helping to bring the particle and gravity communities together.

Particles in this theory are rising due to excitations of a string, but included in these particles are one with zero mass and two units of spin. If a quantum theory of gravity existed, then the particle carrying gravitational force would include the zero mass and two units of spin. This particle has been called a graviton. Early string theorists proposed that their theory shouldn't be applied to hadronic particles, but to quantum gravity.

In string theory, the strings are colliding in a small and finite distance. The zero distance behavior means that scientists can now combine gravity and quantum mechanics. This theory allows scientists to talk about string excitation carrying gravitational force.

The string theory not only helped overcome some hurdles within physics, but it also inspired young people to learn this complex math to study quantum theory of interacting strings. So how can we describes these strings?

It's all in a guitar string. One that was tuned through stretching the string and providing tension. When it's plucked, it produces a variety of musical notes, or excitation modes. Put elementary particles in the place of musical notes, and we have the excitation nodes of elementary strings. The string must be stretched with tension to get them into an excited state. But these strings aren't tied to a guitar, but float in space. Still they have tension. But in order for theory to work with quantum gravity, the average string should be near the length scale, otherwise called the Planck Length.

However, these strings are so small, it would be impossible to see them with current technologies. So string theorists have come up with alternatives methods. Classifying string theories by whether or not strings need to be closed loops or if a particle spectrum includes fermions. To include fermions, a special kind of symmetry must be in place, called supersymmetry. This means that every boson has a corresponding fermion, basically particles that transmit forces are related to particles making up matter. To date, supersymmetric particles haven't been observed, but scientists believe these particles aren't able to be detected at with current technology. Still scientists have high hopes for the particle accelerators of the future. Evidence of supersymmetry would be evidence of string theory's ability to be a solid mathematical model for nature, even at the smallest scales.

How do you build a string theory? Start with a tiny, wiggling string, then decide if it should be opened or closed. Then add which you want, bosons only or a combination of bosons and fermions. If a scientist choose only bosons, then they get a bosonic string theory. But if you want to add matter in the form of fermions, then supersymmetry is needed to equally match up the bosons and fermions. A superstring theory is one that combines string theory with supersymmetry. The other question is if quantum mechanics can be included reasonably. With bosonic strings, this can be affirmative if space time dimensions are numbered around 26, while superstrings can be brought down to 10. Any further down is currently beyond these theories, well for now any way.

Below are six string theories and some explanation about how they work.

| Type | Space Time Dimensions | Details |
|---|---|---|
| Bosonic | 26 | Only bosons, so only force, no matter, regardless of whether the strings are opened or closed. Major flaw is particles with imaginary mass, otherwise known as tachyon. |
| I | 10 | Supersymmetry includes forces and matters, regardless open or closed strings, no tachyon, but their group symmetry is 32. |
| IIA | 10 | Supersymmetry between forces and matter, but only closed strings, no tachyon, massless fermions are massless and nonchiral |
| IIB | 10 | Supersymmetry between forces and matter, only closed strings, no tachyon, massless fermions are chiral, only spinning one way |
| HO | 10 | Supersymmetry between forces and matter, closed strings, no tachyon, heterotic or right/left moving strings differ, plus group symmetry is 32. |
| HE | 10 | Supersymmetry between forces and matter, closed strings, no tachyon, heterotic, group symmetry is **$E_8 \times E_8$** |

With all of these theories, one needs to define what they are looking for, specifically the movement of the string and what is involved. Now we can move toward something that is infinitely more fun, because it appeals to the sci-fi junkie in all of us. Yes, it's time to talk about hidden dimensions.

# Chapter 14 – Hidden Dimensions

Hidden dimensions are those dimensions that step outside of the classically defined ones to work with various theories, including superstring theory. So what are the defined dimensions? Let's start with spatial dimensions. Classic physics describes three physical dimensions, easily defined by our own movements. We can move forward/backward, up/down or left/right. No matter what movement we make, direction can be expressed by these three. So if we move up and forward, then we are moving in a linear combination.

To describe this in terms of dimensions, a line would be one dimension, a plane would be two and a cube would be a three dimensional object. These are what spatial dimensions really are, the dimension defined by the space in which we move.

Another dimension is time. One dimension of time is a temporal dimension, a way to measure physical change. It is perceived as different from the spatial dimensions because there is only one and we are subject to moving in time's one direction.

Physics equations do not model reality and time the same way that humans perceive it. In fact, these equations are symmetric with respect to time and continue to be so if both time and other quantities have been reversed. The perception of time running in one direction is an artifact as part of the laws of thermodynamics. When time and space are treated as components of a four dimensional manifold, otherwise known as space-time.

These four dimensions are considered the accepted norm in physics, but other theories attempt to unify these four dimensions with other ones. We discussed superstring theory and noted that many of those theories required at least 10 space-time dimensions. However, as of this date, no observational or experimental data has been produced to confirm the existence of other dimensions. Many scientists argue that if these dimensions exist, they may be too small for our current technology to find in the current experiments. The future could change that, as detectors and other measuring tools are refined even further.

In addition to tiny and curled up dimensions, there may be extra dimensions not obvious because the matter related with the universe we can observe is localized to the four dimensional subspace. So the extra dimensions might not be small, but could be extra-large instead. D-branes are objects that are dynamically extended within various dimensionalities predicted by string theory play a role. These dimensions may have open string excitations, associated with gauge interactions, thus confined to brane at their ends. Closed string mediating the gravitational interaction are free to propagate into spacetime. There may be some relation to why gravity dilutes itself as in builds into a higher dimensional size. This may be referred to as brane physics.

Extra dimensions can be universal if all the fields involved are equally free to promulgate within them. At the same time, dimension can be comprehensive for networks embedded space and can be characterized by their spatial constraints. While the debates within quantum physics will continue as scientists refine theories and experiments, none have had such a profound effect on the field of quantum mechanics like the Bohr-Einstein debates.

# Chapter 15 –
# Bohr-Einstein Debates

The Bohr–Einstein debates is a succession of public debates on quantum mechanics that involved Albert Einstein and Niels Bohr. The importance of these debates is based on the philosophy of science and how it expanded the views of many scientists. These debates occurred at a time of great change and discovery within quantum physics. While Bohr was judged the victor, it establish fundamental character of quantum measurement but scientific consensus isn't necessarily been achieved.

During the 1920s, the quantum revolution occurred under the direction of both of these scientists and their debates were about understanding the changes. The uncertainty principle and various probability equations were interpretations that Einstein was unwilling to accept.

His refusal to accept the upheaval as complete, reflected his personal desire to have a model that explained the underlying causes for the results these experiments were producing. He though there was more out there to be discovered and didn't want it to get swept under the rug of the uncertainty principle. Bohr didn't have these issues, but had made peace with the apparent contradictions one could find in quantum mechanics. But Einstein moved his position over time and his contributions to the field cannot be denied.

During what is referred to as the post revolution period, Einstein moved through various stages that allowed him to modify his position. During his first volley, Einstein used some ingenious thought experiments to challenge the principle of indeterminacy, which he felt could be violated. His first volley was the orthodox conception of electrons and photons, during an international conference in 1927.

At this argument, Einstein argues that incident particles have velocities in a practical sense that are perpendicular and only the interaction with this a deflection screen can change the original direction of propagation. A conservation of impulse implies that the sum of these impulses the interact will be conserved but if the incident particle is deviated at the top, the screen recoils toward the bottom and the reverse is also true.

Using realistic conditions, the mass of the screen remains stationary, but it is possible to measure even an infinitesimal recoil. The interference happens precisely because the system's state is the superposition of two other states whose non-zero wave functions are near one of the two potential slits. On the other hand, if the particle passes only through one slit, then the systems are a statistical blend of the two states, which makes interference impossible. If Einstein is correct, then damage to the indeterminacy principle would exist.

Bohr's answer was to explain Einstein's idea more plainly but using a screen that was mobile versus fixed. Bohr observed very precise knowledge of the screen's vertical motion was a vital supposition in Einstein's argument. However, Bohr believed, an extremely precise determination of the screen's velocity, when applied to the indeterminacy principle, suggests an unavoidable imprecision of its position. Before the process started, the screen would occupy an indeterminate position, to some extent.

The ideal experiment must average over any and all of the possible positions the screen could occupy and for every position, there would be a fixed point and a different type of interference, creating the perfectly destruction into the perfectly constructive. But again, our attempt destroys the possibility of interference.

Any additional experimental apparatus was considered part of the experiment as it could introduce new effects of potential interference, which can influence the end results. Bohr believed that to illustrate the microscopic effects, one needs to bring these experiments into the macroscopic through the use of various apparatuses. This continues to be a difficulty that is called the measurement problem.

Einstein's next criticism was supposed to prove a violation occurred of the indeterminacy relation between time and energy. So he tried to use the thought experiment designed by Bohr in 1930. Mr. Einstein expressed an idea that the existence of Bohr's experimental apparatus could be used to emphasize essential elements and key points.

Using a box with electromagnetic radiation, a clock that controls the shutter's opening, and his famous $E = mc^2$, Einstein attempted to show that in principle the mass of the box with electromagnetic radiation can be determined. In addition the energy within the box can be measured or determined with a precision that makes the final product less than what is implied by the indeterminacy principle.

But Bohr didn't give up. Instead he proved that Einstein's subtle argument couldn't be conclusive, but he did so using an idea from Einstein, that of the equivalence between gravitational mass and inertial mass. Basically, the box would have to be up in the air on a spring in the middle of a gravitational field and the weight would have to be obtained through a pointer attached to the box that links to a scale index. The unavoidable uncertainty of the box's position translated into uncertainty about the pointer's position and the weight plus energy determination. Thus the entire thought experiment only seemed to demonstrate the uncertainty principle even more clearly. This back and forth continued into what is known as the second stage of their debate.

The second phase was regarding the orthodox interpretation characterized by the fact that it is impossible to concurrently define the values of certain discordant quantities. Einstein believed there was an ability to measure these values and so a line of research into hidden variables was done to make quantum physics complete from Einstein's point of view.

The third stage focused on the EPR paradox is discussed in greater detail in Chapter 7. This argument continued well into the 1950s and included many other scientists who discussed quantum entanglement as it relates to particles. Bohr responded to Einstein's EPR paradox, particularly questioning the expression that one could complete the experiment without disturbing the system in any way. With the use of possibilities and other theories of interaction, Bohr combated Einstein's paradox.

In the fourth stage of their debate, Einstein continued to refine his position, stating that quantum theory disturbed him because of its total renunciation of all minimal realism standards, even at the microscopic level. To this day, the understanding is still not complete and scientists continue to debate without a consensus on determinism.

# Chapter 16 –
# Physics in the Real World

While physics does play a role in our lives, most of it involves things we don't really think about. For example, physics helps to define how our world is put together on the molecular level. Understanding that helped them to split atoms and use various waves to transmit information via data and sound.

At the same time, it's interesting to look at how brane physics, just one area, can be used to help us understand dimensions, even ones that might not be easily found or seen. Various aspects of brane physics have been used in cosmology. For example, brane gas cosmology attempts to clarify why the three dimensions of space by topological and thermodynamic contemplations.

This idea put forth three as the largest number of spatial dimensions, because that is where strings can generically interconnect. Initially, there might be multiple windings of strings around dense dimensions, but space only expand to macroscopic sizes but only once these windings are removed. To remove these windings, oppositely wound strings must find each other and then annihilate. But they can't find each other to annihilate in only three dimensions, so this follows the idea that only three dimensions of space are able to grow larger with this initial configuration.

With just this one aspect, we have seen how quantum physics is working to understand how dimensions, space and time work together in our universe. It is the greater understanding of how our universe works that ultimately brings quantum physics or mechanics to us.

# Conclusion

As we have seen, Quantum Physics is built on observations of the behavior of matter and energy. But it also involves taking those observations and creating mathematical equations to explain them. This is a science where observation is critical. While constants have been agreed upon, with the invention of better and more precise measuring tools, Quantum Physics continues to refine its theories.

Exploration of the molecular level of the world takes some degree of faith, because most of these theories are still just that, with evidence for or against it just another experiment away. Scientists in the field often have disagreements about how and if their equations are properly mapping what they are observing. Einstein disagreed with many scientists of his time and his own theories have continued to be put to the test.

Yet, as with any science, experimentation and hypothesis continue to rule the day. Quantum Physics grows as a scientific field of study because each new generation of scientists is willing to go one step further in their study of the molecular world. These attempts to define our Earth and Universe only add to our collective store of knowledge.

It's important to remember that the mathematical equations involved in Quantum Physics are complex and based on agreed upon constants. So scientists are also testing those constants within their experiments. In Quantum Physics, nothing is absolute but everything is open to a better interpretation and understanding.

Throughout these chapters, we have looked at several experiments and theories that make up the field of quantum physics. These theories and experiments have become part of the foundation of quantum mechanics and over time, scientists and physicists have continued to use these ideas to refine their understanding of how the universe works on a microscopic level.

This information has assisted in the understanding of how stars are born, what matter and force do when they interact with each other on a particle level and also in larger masses.

Learning about protons, electrons and the radiation used to measure them and their movement has become part of our collective understanding of radiation itself and even how energy is created and stored. Yet there are still so many things that quantum mechanics can't explain. For instance, how particles and waves can both be part of the movement of light, as they have described in wave particle duality. Truly being able to define the contradictions inherent in these theories is the ongoing work of physicists and scientists.

As technology improves, these same scientists may eventually be able to create the experiments that allow them to find the answers to the questions still out there.

\*\*\*\*\*\*\*\*\*\*\*\*\*\*\*\*BONUS\*\*\*\*\*\*\*\*\*\*\*\*\*\*\*\*

# Astronomy

## A Beginners' Guide to Space

*Black Holes, New Galaxies, and more*

# Introduction:

On any given night, one can go outside and see a magnificent show in the sky. Stars, planets, comets and a host of other offerings are visible. But what are we looking at? Taking a tour of the universe, a beginner can receive an overview of the wonders of space and the night sky. When children create a tower with blocks, no two blocks or towers are exactly alike. Space is much the same way. Each part of the universe is unique but a necessary part of the whole.

Throughout the following chapters, we will explore the universe, starting with one of the largest blocks, galaxies. Each different block will assist us in understanding the other blocks or pieces of the universe. These blocks interact with each other and those relationships influence the formation of galaxies, stars and planets. The science behind these relationships has grown over the years, assisting humans in understanding the universe and what it's made of. Still as scientists find out more, additional questions are raised, creating ongoing lines of study and exploration. This book doesn't cover every area of space, but provides a simple foundation to begin your own exploration of the night sky in your backyard. So sit back and enjoy this brief tour of space, the never-ending frontier.

# Chapter One – Space: What's Really Out There?

On a clear night, anyone can step out in their backyard and see a magnificent show in the sky. Hundreds of thousands of stars, planets and comets grace the night sky. Standing outside, it can be overwhelming to try and process all the beautiful things you see. But what are you really looking at?

Space is best described using the example of building blocks. Each part of space is made of other parts, building from one star to thousands of galaxies. By learning about each block, you can build an understanding of what the whole tower looks like, even if we can't see the whole universe from our back yards.

So what are the parts or blocks we will learn about? First, we'll discuss galaxies. It's important to know what they are and the various types, as these are not only the homes of stars, but their birthplaces. Solar systems are the block within the galaxies. Within solar systems are the stars and planets, another critical block. With millions of stars visible from Earth, constellations provide a way to group and describe various star formations, while providing a natural GPS.

The travelers within our universe are comets, streaking across the sky from one galaxy to another. Other unique events, such as the northern lights, occur without warning, but provide wonderful decoration to our night sky. Let's start our journey through space with our first block, galaxies.

# Chapter Two –

# Galaxies

Galaxies themselves are made of gas, dust and numerous stars, held together by means of gravity. This building block can be particularly hard to fathom, because the visible universe is said to contain at least 100 billion galaxies. That's a lot of dust, gas and stars. Small galaxies have under a billion stars. A bit mindboggling, isn't it?

All galaxies are classified into three types: spiral, elliptical and irregular. A spiral galaxy takes on the appearance of a flat disc, similar to a Frisbee. The disc is made up of a bulge in the center and arms extending outward from that center. The stars, dust, planets and gases rotate around this center in regular intervals, but at such speeds that it comes to resemble a pin-wheel. These types of galaxies are also known for producing new stars, due to the abundance of gas and dust that fuel star formations.

Elliptical galaxies lack the arms of the spiral cousins. Instead they tend to take on the shape of a long cigar or extremely large circles. Because these galaxies have less dust and gas, they tend to produce less new stars. These galaxies also more likely to be older and past their star bearing years, so to speak. While their stars rotate around the galactic center, as spiral galaxies do, the directions of rotation appear more random. Some of the largest galaxies are elliptical. Smaller elliptical galaxies are referred to as dwarf elliptical galaxies.

In this case, size is a relative term, because both types include millions, if not billions of stars.

Of course, not every galaxy fits neatly into one of these categories. As a result, we have our third type of galaxy, known as the irregular galaxy. The shape of these galaxies appears to be influenced in part by the larger galaxies that make up its neighbors. Irregular galaxies often appear misshapen or lacking a true form.

As irregular ones demonstrate, galaxies can and do influence each other. Typically, galaxies tend to group together, although they can be found alone or in pairs. Occasionally, two galaxies collide or merge with one another, and the influx of dust and gas contribute to an increase in star formation. As galaxies group together, their gravitation influence can take an irregular galaxy into a spiral, and eventually age it into an elliptical. As a family of sorts, these groupings influence the speed of star formation and overall shape of their neighbors.

Within these galaxies, one finds another of our blocks, the solar system. A solar system is comprised of a central star, with planets that orbit around it. The planets themselves may have moons orbiting around them. All of these pieces are in continual motion. This constant motion might make you wonder why we don't all suffer from motion sickness. Thankfully, gravity and a host of other natural laws help us to feel as if we are standing still during this continually spinning ride.

Our solar system is made up of the Sun, planets, moons and asteroids. These planets, moons and asteroids, as well as other things such as comets, are drawn to their Sun through its gravitational pull and so they remain in their various orbits. The Sun creates energy through nuclear fusion of hydrogen into helium, which expels electromagnetic energy that peaks as visible light. Within our solar system, the Earth is the third planet from the Sun. Its position within the solar system means the Earth receives just the right amount of the Sun's energy. Too little, and it would be too cold to sustain life, but too much energy from the Sun and the temperature would be too hot to sustain life.

Imagine for a moment the warm sticky air of a humid climate. When the temperature soars, it can be hard to breathe and draining to move. Imagine if the temperature was twice as high, or even three times. There is no way that people and animals, if any, could survive those temperatures. Yet the amazing positioning of the Earth keeps us at just the right temperature to sustain life. With a slight tilt, the Earth is also able to sustain seasons. Many of us who survive the northern climates extreme cold and harsh winds during the winter have learned to value those humid summers, taking the opportunity to thaw out.

Other planets in the solar system cannot sustain life, not only because of the differences in temperature, but also due to a lack of the elements and atmospheric mix necessary to sustain life. The four planets closest to the Sun (Mercury, Venus, Earth, and Mars) are referred to as the **terrestrial planets** because of their solid and rocky surfaces. The four large planets beyond the orbit of Mars (Jupiter, Saturn, Uranus, and Neptune) are known as **gas giants; large, low density planets made up primarily of hydrogen, helium, methane and ammonia in a gaseous or liquid state**. Tiny, distant, Pluto has a solid but icier surface. While it is closest to the Sun in its orbit, Pluto has a thin atmosphere, which then collapses to the surface when it is furthest from the Sun. As a result, for a period of its orbit, Pluto behaves as a comet.

Our solar system is not the only one with planets. As we move on to examine the next block in our tower of the universe, we'll find out more about the variety of planets making up their part.

# Chapter Three –

# Planets

The first thing that comes to mind when we think of planets is our own little solar system, but most often, we think of our own planet, Earth. All planets can be classified by their atmosphere, the elements that make up the planet, its temperature, its position in relation to its parent star and even by the satellites it does or doesn't have. While there are a large variety of planet types, we are going to examine at just three types here. These examples are just a taste of the amazing diversity of planets within our universe.

Chthonian Planet

A chthonian planet is a class of celestial objects resulting from the stripping away of a gas giant's hydrogen and helium atmosphere, along with the outer layers of the surface of the planet. This typically happens when a gas giant is in close proximity to a star, through a process called hydrodynamic escape. During this process, heavier atomic molecules escape into space as a result of repeated collisions with lighter atomic molecules. Its bumper cars for atomic molecules on an immense scale. Eventually, the planet is simply left with a core, as it continues to orbit around its Sun.

Goldilocks Planet

A goldilocks planet typically refers to a planet in the habitable or "just right" zone from their parent star to sustain life as we humans understand it. A planet needs to be in a unique position in relation to their parent star, neither too close nor too far to definitively rule out the possibility of liquid water on the planet's surface. However, it is important to note that planets, such as gas giants, may be within the habitable zone, yet unlikely to host life. Obviously, we live on the best example of a goldilocks planet. Scientists are especially interested in these types of planets, both for the possibility of intelligent life besides humans or as potential new homes for humans.

An example of the search for goldilocks planets is the Kepler Mission, a NASA project using existing technology to examine space for these particular planets within these defined habitable zones. It is estimated that at least 11 billion of these planets may exist, each orbiting their own star. The nearest one is said to be 12 light years away, but we as yet have no way to travel to this potential goldilocks planet.

Lava Planets

Lava planets orbit close to their parent star, but with an eccentric orbit. As a result, the gravity from the star would distort the planet, creating friction that produces internal heat. This heat could melt rocks into magma, which would erupt through volcano like structures. These planets may resemble Io (a system moon that orbits Jupiter) with its extensive sulfur concentrations on its surface, often associated with continuous volcanic activity. In addition, the intense stellar irradiation from their close orbit to their parent star could melt the surface crust directly into lava.

This intense stellar irradiation means that the illuminated surface of these planets could be covered in more or less a lava ocean, while the dark side may have rock rain (caused by the condensation of vaporized rock from the hotter side) or lava lakes. The mass of the planet also plays a role. Plate tectonics on terrestrial planets are related to planetary mass, so planets with more mass than Earth are expected to exhibit plate tectonics and as a result, intense volcanic activity. Lava planets can also occur temporarily due to a large impact, such as the collision with Earth that formed our moon.

As you can tell, the planets offer a wide variety of blocks, contributing a never ending stream of possibility and exploration to our universal tower. At the same time, stars have continually played a part in each of these blocks. So what is the life cycle of a star? Let's find out.

# Chapter Four –

# Stars

As we discovered in our exploration of galaxies, a star's beginning involves dust and gas. Within the galaxies are dust clouds filled with dust and gases. Turbulence or collisions deep within these clouds allow knots to form with sufficient mass to permit the gas and dust to collapse under its own gravitational attraction. As the cloud collapses, the material in the interior commences warming up. This is called a protostar, because this hot core of the collapsing cloud will one day be an illuminating star in the night sky. However, these clouds can also break up into two or three different cores, creating several stars. Scientists believe this could account for why stars, particularly those in the Milky Way, are paired or have groups with multiple stars.

As the collapse continues, the gravitational pull of the core draws in additional dust and gases. This material may become part of the star, or turn into the planets, asteroids or comets that circle the star. Some of this material will simply remain dust floating in orbit around the newly formed star.

Main Sequence

Eventually this baby star will mature to the main sequence or star adulthood of its life cycle. Scientists estimate that a star the size of our Sun takes approximately 50 million years to reach its main sequence stage. But what keeps a star shining over millions and in some cases billions of years? Stars are fueled by the constant nuclear fusion of hydrogen to create helium deep within the star's interior. The central regions within the star provide a steady outflow of energy from this process, enough to keep the star from collapsing under its own weight, while giving the star its signature shine.

Main sequence stars traverse a wide assortment of luminosities and colors, and can be classified according to multiple characteristics. The smallest stars, classified as red dwarfs, possibly contain a small mass, as minimal as ten percent of the mass of the Sun, and emit only 0.01% of its energy, glowing faintly at temperatures between 3000-4000K. Despite their diminutive nature, red dwarfs are the most numerous stars in the Universe, with an impressive lifespan of tens of billions of years.

On the other end of the spectrum, massive stars, known as hypergiants, may be a hundred or more times beyond the mass of the Sun, with surface temperatures that reach in excess of 30,000K. Hypergiants emit hundreds of thousands of times more energy than our Sun, yet their lifespan is significantly shorter, often only a few million years. These types of stars are considered extremely rare and the Milky Way galaxy has been documented as having only a handful of these hypergiants.

The Star's Twilight

Scientists have determined that the larger the star, the shorter their lifespan. But all stars eventually come to the stage in their development where the hydrogen has all been converted to helium and the nuclear reactions cease. Deprived of its necessary energy production, the core of the star begins to collapse and becomes increasingly hotter. Since hydrogen is still available outside the core, hydrogen fusion will continue in a shell surrounding the core. In the meantime, the hot core pushes this shell and the other outer layers of the star outward, creating an expansion and cooling effect that transforms the star into a red giant.

Again, we come back to the size of the star's core. A sufficiently massive star may have its core become hot enough to support more exotic nuclear reactions, consuming the helium within its core and producing a variety of heavier elements, all the way up to iron. While this may temporarily extend the life of the star, it is but a short reprieve. Over time, the star's internal exotic nuclear fires become increasingly unstable - sometimes burning furiously, other times dying down. These unstable nuclear reactions cause the star to pulsate and throw off its outer layers, gradually enshrouding the star's core in a cocoon of gas and dust. This gas and dust play an important part throughout a star's life. But what happens next depending entirely on the size of the star's core.

White Dwarfs

A white dwarf is created when an average star, similar to the Sun, ejects its outer layers until the stellar core is exposed. This white hot cinder, while technically a dead star, is roughly the size of the Earth. It was puzzling why these masses didn't collapse further. Through the use of quantum mechanics, scientists found the answer. Pressure from fast moving electrons keeps these stars from collapsing further. The more massive the core of the star, the denser the white dwarf will be. Conversely, the smaller a white dwarf is in diameter, the larger its mass will be.

These stars with their contradictory ending are very common. Our Sun will be a white dwarf at some point in the distant future. Because these white dwarfs are no longer producing their own energy, they are fainter and eventually will fade away as their cores continue to cool. But stars with a mass of up to 1.4 times our Sun will have a different fate, as the pressure from the electrons can't stop the continued collapse of the core. Is their demise more eventful?

Nova

White dwarves can form in a binary or multiple star system. When they do, their end may come in the form of a nova, the Latin word for new. While originally scientists thought that novae were the creation of new stars, they are in fact white dwarves or very old stars. These very old stars, if close enough to a companion star and with significant mass to cause a great enough gravitational pull, will drag matter from that companion star (primarily hydrogen). The result is the creation of a new surface layer, where nuclear fusion can occur again, abet temporarily. The star then brightens for the interim, before expelling all the excess material it has accumulated and then the process begins again. These smaller surface explosions are the novae. The larger stars will eventually accumulate so much matter that they collapse completely and explode. These are called supernovas.

Supernova

As we have seen, the universe provides a wonderful show using the various blocks it has available. Yet a supernova is an amazing phenomenon all on its own.

Main sequence stars over eight solar masses die in a titanic explosion, the incredible supernova. While a nova's explosion occur on the surface of the star, a supernova is the result of the star's core collapsing and then exploding. Massive stars produce iron through a complex series of nuclear reactions. Once this is achieved, the star has literally wrung all the energy it can out of nuclear fusion, because fusion reactions that produce elements heavier than iron require energy and do not produce it. Without any way to support its own mass, the iron core of the star collapses. In a matter of seconds, the core shrinks from approximately 5000 miles across to just a dozen miles and the temperature can spike to over a 100 billion degrees. The surface and outer layers of the star initially begin to collapse inward toward the core, but with the sudden spike in temperature and the massive release of energy, these layers are thrown violently outward. This intense release of energy means a supernova can outshine a galaxy for days or even weeks. All naturally occurring elements and a fertile array of subatomic particles are fashioned during these amazing explosions. On average, a supernova occurs about once every hundred years in the typical galaxy.

Neutron Stars

Again, we have to return to the size of star's core. While a smaller star becomes a white dwarf and a larger star can turn supernova, stars in the middle of the spectrum have a unique destiny as well. This block in our tower is called a neutron star. This phenomenon occurs for those cores at the center of a supernova that contain between about 1.4 and 3 solar masses. The collapse of the core continues until electrons and protons combine to form neutrons, creating a neutron star. These stars are incredibly dense, similar to an atomic nucleus. The gravitation at the surface of a neutron star immense, due to the amount of mass packed into such a small volume. As a white dwarf will strip hydrogen from other nearby stars, a neutron star in a multiple star system can accrete gas by taking it from nearby stars. This type of star can also be known as a pulsar, due to the radiation beams produced by its magnetic poles. If the star is oriented to Earth, as the beam sweeps by it appears to pulse, hence the name pulsar.

So what happens to collapsed cores larger than 3 solar masses? What happens next isn't a block, so much as a hole.

Black Holes

Black holes have always been fascinating, in large part because of the unknown aspect they present. Scientists have discovered that these black holes are a collapsed core of a star. They collapse so completely and are so dense that their gravitational pull sucks in everything, not even allowing light to escape. They are detected indirectly by observing the gamma rays and x-rays omitted as matter is being pulled into the black hole, as current instruments are not able to detect them directly.

From Dust and Gas…

The dust and debris left over from novae and supernovae blend into the surrounding gas and dust, enhancing it with heavy elements and chemical compounds. These materials are recycled and become the building blocks of the next generation of stars and their planets as they orbit through their home galaxy.

# Chapter Five –

# Constellations

Now that we have a better understanding of the life of a star, it's time to focus on how stars have been used for centuries as part of navigational and seasonal system. Constellations, groupings of stars, are used as memory aids to help individuals locate various stars in the sky, making the night sky easier to explore. Let's face it, the night sky is huge and it can be hard to find individual stars without a map of sorts. Constellations provide a natural map, breaking the night sky into more manageable bites.

Today, looking for constellations can be a fun activity on a camping trip or during an evening the backyard, but for many farmers and travelers on land and sea, those constellations have guided them for generations.

Originally, farmers used constellations to assist them in knowing when it was time to plant and when it was time to harvest. Why was this? In part because areas of the earth didn't see dramatic season changes. But certain constellations only appeared in specific seasons. Hence, if a specific constellation was in evidence, it was spring and time to plant.

Stories were created to go along with the constellations, in part as memory aids, but they also were entertaining. Remember, back then, they didn't have television, the Internet or video games. Over time, these constellations were used by sailors to orient themselves in the ocean. Pole stars, those that stayed visible throughout the night, were used as constants to define direction. Sailors would orient based on those pole stars and then using other star groupings to sail in the right direction. Today, stars and their constellations are still used to determine location based on measurements that correlate to the latitude and longitude here on earth.

Throughout history, constellations have been used by a variety of cultures, giving some of these groupings common names, such as Orion and his belt, the Big Dipper and the Little Dipper. What is important to remember is that while these stars may appear close together to us here on earth, but in reality, they are light years apart and moving in their own unique orbits.

Today, the International Astronomy Union (IAU) is an official international body of professional astronomers, who assign official designations to celestial bodies. They currently have a list of 88 official constellations or areas. When they say a star is in a constellation, the IAU means that it is within the boundaries of a certain area of the sky. These 88 areas cover the entire sky as seen from Earth.

This block of our night sky is best understood as a means for humans to comprehend all the other blocks we have discussed.

# Chapter Six –

# Northern Lights and Comets

These blocks are the unique, providing a finishing touch to our discussion of the parts of the universe. Comets are small, fragile and irregularly shaped pieces of matter, often composed of non-volatile grains and frozen gases. These bodies follow elongated paths around the Sun. Comets themselves are cold objects and we can only see them thanks to the gases in their comae and tails fluoresce in sunlight. How are the comae created? Simply put, when a comet nears the Sun, the radiation from the Sun starts subliming the volatile gases, blowing away small bits of solid material. This material expands into an enormous escaping atmosphere creating the coma (comae). Comets are gravitationally bound to the Sun and are regular neighbors within a solar system.

Scientists believe that comets are constructed of leftover debris that didn't become part of the planets. As a result, this material is considered a snapshot of the earliest period of the solar system and a way to learn more about our little corner of the universe.

The nucleus of a comet contains silicates akin to some ordinary Earth rocks in composition, and are glued together into larger pieces by the frozen gases. One of these nucleus looks as if it includes complex carbon compounds and some free carbon, making it very black in color. A young comet will contain more frozen gases, including ordinary water. In space, water behaves similar to dry ice, going directly from solid to liquid. Overtime, a comet will lose most its ices (such as water, carbon monoxide, methane, ammonia and formaldehyde), becoming a fragile old rock in appearance, nearly indistinguishable from an asteroid.

A comet has a weak gravitational pull, so the escaping gases and solid particles that form their tails never fall back to the nucleus. Radiation pressure from sunlight forces the dust particles back into a tail in the opposite direction of the Sun. The gas molecules are torn apart by the ultraviolet light, becoming ions that form the ion tail, which becomes fluoresce in sunlight. While a comet may appear to have only one tail, it in fact has two. Depending on the comet, only one or neither tail may be visible.

Northern Lights or aurora borealis are a phenomenon created by the Sun. The aurora begins on the surface of the sun, as solar activity ejects a cloud of gas or a coronal mass ejection (CME). When one of these reaches the Earth, it collides with our magnetic field and creates complex changes to happen in the magnetic tail region. As a result, currents of charged particles are crafted. As they flow along the lines of magnetic force to the Polar Regions, colliding with oxygen and nitrogen, the dazzling aurora light becomes visible.

These Northern Lights are simply another amazing event adding to the beauty of our skies. So do you need to be an astronomer to enjoy it?

# Chapter Seven –

# Enjoying the Beauty in the Skies

As we have examined these various blocks in our universe, we can have a great appreciation for the magnitude of the chemical and physical changes that must occur to create the stars, planets and so much more. But you don't need to be an astronomer to enjoy the beauty of the skies and the wonders of space. In fact, all you really need is to look up at the sky wherever you are.

Providing assistance to stargazers are beginner guides to the constellations, often available at your local library, to teach you their identifying stars. Using one of these guides and your own backyard, it can be fun to find the various constellations. With an inexpensive telescope, one can get an even better view of the stars in the sky. But keep in mind the stars are being viewed through our atmosphere, causing a slightly hazy effect. This is why pictures taken with space telescopes appear much clearer, as they don't have the same interference.

As a family, research the stories behind various constellations, even telling them as you stargaze. News stations will often mention if a comet, planet or other event will be visible from your home. If you can, take the time to go outside and look for it.

Planetariums are another way to explore the stars. Presentations make visible the stars and groupings that might not be seen from your home hemisphere. In addition, these presentations can be made using different viewpoints, including space, showing the universe without an earthbound view.

By far and away, the night sky provides beauty, but it also a source of ongoing exploration, as we learn more about these blocks making up our universe, our galaxy and our solar system. Happy stargazing!

# Conclusion:

As we have realized over hundreds of years, space is filled with a variety of phenomenon. The timing of these events is in terms of millions and billions of years, far exceeding the average human life span. The creation of one star begins long before we are born and will not end for centuries after we have passed away. Planets that seem solid also change in due course as the stars surrounding them influence their surfaces and atmospheres due to their gravitational pulls and the nuclear fusion reactions. Yet the universe made available many clues to understand how it builds these stars and planets, even if we personally will not live through an entire life cycle of a star or planet. On the hunt for these clues, scientists continue to study the universe, accessing supplementary data to further explain how it was created.

In the meantime, there is plenty of information available at our fingertips to give an in-depth study of any particular area of space. Resources available include NASA, the IAU and other scientific research bodies. Hopefully, this overview has given you the incentive to dig deeper and learn more about your favorite constellation or star. Astronomy has always been a part of the human condition, as we try to learn more about the space surrounding our floating home. Look up!

Printed in Great Britain
by Amazon